THE SUN

THE SUN

LEROY VINCENT

CONTENTS

CHAPTER 1

Introduction to the Sun

Among the billions of stars in the vast expanse of the Milky Way Galaxy, there is one star that, while seemingly unremarkable in the grand cosmic scale, holds unparalleled significance for us. This star is none other than the Sun. Its importance goes beyond being the subject of a key unit in the Australian Science Curriculum; the Sun is fundamental to understanding the Universe and our place within it. Everything we observe and experience on Earth is influenced by this closest star, making it a perfect starting point for our exploration.

The Sun is often described as a massive ball of "light" and "heat," but this description only scratches the surface. In reality, the Sun is a colossal sphere of plasma whose gravitational influence dominates the Solar System, accounting for 99.8% of its total mass. If the Sun were hollow, it could contain about 1.3 million Earths. Its structure, defined by layers that range from the core to the outer corona, is remarkably complex and efficient, much like a nuclear power station operating on a scale beyond human comprehension.

In this chapter, we will delve into the Sun's role as a star, understanding its properties as a "Main Sequence Star," and examining its major elements. We will then explore its critical role in the Solar System and ponder the implications of life without its vital heat and light.

Importance of Studying the Sun

Studying the Sun in great detail is essential for several reasons:

- **Understanding Other Stars**: As the nearest star, the Sun provides the best opportunity to study stellar characteristics. This knowledge helps us understand the billions of other stars in the universe and our cosmic neighborhood.
- **Earth's Habitability**: The Sun is the primary driver of Earth's climate and weather systems. Understanding its variations enhances our comprehension of many aspects of the solar system and natural phenomena on Earth.
- **Advancements in Science**: The Sun's processes, from its core to its corona and solar wind, contribute to advancements in atomic spectroscopy, nuclear science, and plasma physics. This knowledge is vital for various scientific fields.
- **Practical Applications**: Solar research has practical applications, particularly in space weather. Improved space weather predictions enhance the reliability of the Global Positioning System (GPS), communications, and other space-based systems, which are now integral to the global economy.
- **Defense Against Solar Eruptions**: Understanding and predicting geomagnetic storms caused by severe solar eruptions can have significant economic and social benefits. These advances could also improve early warning systems for near-Earth asteroid impacts, protecting our planet from potential threats.
- **Climate and Global Warming**: The Sun's radiation affects Earth's climate and weather systems. Solar research is crucial

for understanding and mitigating the challenges posed by climate change and global warming.

By studying the Sun, we gain invaluable insights into the universe, improve our technological capabilities, and protect our planet. The Sun, despite its seemingly ordinary nature, is a linchpin in the grand tapestry of the cosmos.

CHAPTER 2

The Structure of the Sun

The Sun is a complex astronomical body that plays a crucial role in sustaining life on Earth. As the central star of our Solar System, its influence extends far beyond providing light and warmth. To understand this remarkable star better, we must delve into its intricate structure and the processes that occur within it.

The Sun is a massive sphere with a mean radius of approximately 695,508 kilometers, making it about 109 times the diameter of Earth. Its structure is composed of several distinct layers, each playing a critical role in the Sun's overall function and behavior.

The Corona

The outermost layer of the Sun is the corona, which is characterized by incredibly high temperatures that can exceed millions of degrees Celsius. The corona extends millions of kilometers into space and is visible during solar eclipses as a bright, shimmering halo. The transition from the corona to the underlying layers is marked by a sudden and substantial increase in temperature, a phenomenon that continues to intrigue scientists.

The Inner Structure

Beneath the corona lies the Sun's inner structure, which can be divided into three main regions: the core, the radiative zone, and the convective zone. These regions are separated by distinct boundaries that define their unique physical and dynamic properties.

The Core

At the very center of the Sun lies the core, the powerhouse where the Sun's energy is generated through nuclear fusion. The core's immense pressure and temperature, which reaches around 15 million degrees Celsius, facilitate the proton-proton chain reaction. In this process, hydrogen nuclei (protons) fuse to form helium, releasing a tremendous amount of energy in the form of gamma rays and neutrinos. This energy production is crucial as it sustains the Sun's radiance and supports life on Earth.

The Radiative Zone

Surrounding the core is the radiative zone, where energy is transferred outward through the process of radiative diffusion. In this region, photons are absorbed and re-emitted by particles, gradually making their way towards the outer layers. This slow journey of energy can take thousands of years due to the dense environment, resulting in a gradual outward migration of heat.

The Convective Zone

Above the radiative zone lies the convective zone, characterized by turbulent, convective currents that transport energy to the Sun's surface. In this layer, hot plasma rises towards the surface, cools down, and then sinks back down to be reheated. This convective process creates the granulated appearance of the Sun's surface and plays a critical role in the dynamic behavior of the Sun's outer layers.

The Tachocline

Between the radiative and convective zones lies the tachocline, a thin but significant layer where the Sun's rotation changes from differential in the convective zone to nearly uniform in the radiative zone. This transition region is crucial for the Sun's magnetic field generation and overall solar dynamics.

Understanding the Sun's structure is essential for comprehending the various processes that occur both within and around it. These interconnected layers and their interactions drive the Sun's behavior, influencing everything from solar radiation and space weather to the conditions that support life on Earth.

CHAPTER 3

Solar Features and Phenomena

The Sun, our closest star, measures an astounding 1.4 million kilometers across—about 109 times the diameter of Earth. While astronomy often involves studying celestial phenomena occurring millions or billions of kilometers away, the Sun's relative proximity offers a unique opportunity to observe and understand various solar features and phenomena that directly affect our daily lives. This section provides a comprehensive overview of the many features and phenomena occurring on and around the Sun, garnering interest from a diverse global audience.

These features, despite the diversity of reasons for their study, share a common theme: they are evidence of the Sun's dynamic nature. When discussing solar phenomena, we are delving into the field of solar physics. Astronomers who specialize in the study of the Sun, known as solar astronomers, work within the field of heliophysics. This discipline investigates the Sun's influence on the solar system, particularly on Earth and the Earth-Moon system.

Among the most impactful solar phenomena are coronal mass ejections (CMEs), which can trigger geomagnetic storms on Earth. During periods of peak solar activity, tens of thousands of solar flares can occur, posing "space weather" hazards to astronauts, satellites, and the crews of both short- and long-duration space missions.

Sunspots and Solar Flares

Our Sun is a hive of activity, exhibiting numerous solar phenomena such as sunspots, faculae, flares, CMEs, and variations in solar flow, temperature, and brightness. These phenomena cause fluctuations in the Sun's emissions, altering its brightness and temperature over various timescales, from centuries to the 11-year solar cycle, to shorter intervals during large solar events.

Sunspots

Sunspots are dark, cooler areas on the Sun's surface caused by intense magnetic activity. They appear darker because they are cooler than the surrounding areas, with temperatures dropping by about 1,500 degrees Celsius. These spots can significantly impact the Sun's temperature and brightness by altering the solar plasma's opacity. Sunspots are regions of intense magnetic fields that inhibit convection, leading to reduced surface temperature in those areas.

Sunspots also have high-energy ultraviolet impacts, cooling down the solar surface locally and causing a chain reaction that expands the cooled area. The number of sunspots varies over the solar cycle, peaking during periods of high solar activity.

Solar Flares

Solar flares are sudden, intense bursts of radiation and energy resulting from the release of magnetic energy stored in the Sun's atmosphere. They can release as much energy as a billion megatons of TNT in just a few minutes. Solar flares emit radiation across the entire electromagnetic spectrum, from radio waves to gamma rays, and are often associated with sunspots.

Flares are the solar signs of magnetic activity, transferring energy from the Sun's magnetic system to its atmosphere and releasing high-energy radiation and accelerated solar energetic particles. More powerful flares can also give rise to dense solar winds and strong CMEs. These phenomena can, in turn, interact with Earth's magnetosphere and ionosphere, affecting high-latitude regions where magnetic flux lines intersect the exosphere.

Solar activity continues to have significant effects on humanity and technology. Understanding the causes and impacts of solar variability is crucial for comprehending modern climate change and developing accurate space weather forecasts. These forecasts can help minimize the effects of solar eruptions on terrestrial, airborne, spaceborne, and planetary technological infrastructures.

One of the key contemporary applications of solar research is understanding and forecasting the impacts of solar events on satellite and ground-based power grids and other Earth- and space-based technological systems. Sunspots are often precursors to intensive solar flares, making them essential for predicting space weather events and protecting technological infrastructures.

CHAPTER 4

Solar Energy and Its Impact on Earth

The Sun is the primary source of energy for life on Earth. Solar radiation is a fundamental factor influencing the climate and biogeochemical cycles of our planet. Its practical implications are vast, impacting various aspects of daily life, from ecosystems to technology. Indeed, the presence of water and life on Earth is inseparable from the Sun's heat. Through solar radiation, in combination with Earth's atmosphere and surface, our planet can generate and sustain the remarkably rich and complex world we observe today. Solar radiation significantly affects Earth's energy budget and indirectly influences global climate dynamics, biogeochemical cycles, and hydrological processes. From an environmental and energy perspective, solar energy is one of the most stable and sustainable resources our planet can provide.

One of the best ways to understand the relationship between solar processes and the near-Earth environment is by studying how solar energy reaches Earth's atmosphere and surface. The primary form of solar energy is electromagnetic radiation emitted from the Sun's surface. The importance of solar radiation to daily life is evident, providing essential sunlight for ecosystems and serving as a heat source with considerable economic value. Electromagnetic radiation from the Sun is typically divided into different wavelengths, and the energy incident at the top of Earth's atmosphere is approximately 1361 ± 3 W/m^2. There are mainly two types of solar radiation received by Earth: direct solar radiation and scattered radiation. Although the intensity of solar energy is relatively stable on human timescales, solar activities can show significant variability, with the range of solar radiation variation reaching several percent.

Solar Radiation and Climate

The Sun is the closest star to Earth and the brightest object in our sky. Its interaction with Earth's atmosphere generates phenomena such as the corona, solar wind, flares, and ejections. Much of the electromagnetic radiation and ionized particles from the Sun enter Earth's upper atmosphere, forming an ionosphere that can reflect radio waves. Intense solar radiation exposure can cause hazards and damages to orbiting satellites, power grids, Global Positioning Systems (GPS), human and robotic space missions, and increase air traffic exposure to cosmic radiation. Additionally, the Sun's energy and heat drive weather patterns (through jet streams, weather fronts, and cyclones) and ocean currents.

Due to the Earth's axial tilt relative to its orbit around the Sun, the hemispheres experience varying amounts of light and heat throughout the year. During June and December, one hemisphere experiences the longest daytime and shortest nighttime, while the opposite occurs in the other hemisphere, resulting in predictable seasonal changes in temperature and weather patterns. By interacting with Earth's atmosphere, visible, infrared, ultraviolet, and other forms of solar radiation significantly influence different atmospheric layers. However, not all wavelengths of the Sun's radiation are absorbed equally; some penetrate the atmosphere to reach Earth's surface.

Absorbed solar radiation warms and cools the Earth, depending on whether it is absorbed or emitted, and is ultimately reradiated back into space. This long-wave radiation is partially absorbed by greenhouse gases (CO_2, CH_4, water vapor, etc.) in the atmosphere, causing warming in the lower atmosphere and at Earth's surface. Thus, the Sun exerts a greater warming influence on Earth than ambient temperature cooling. Consequently, variable and uncertain solar and space physics relationships directly affect near-Earth space and human activities. Moreover, the Sun is also an active emitter of cosmic rays, contributing to our understanding of space weather.

Understanding the relationship between solar energy and Earth's climate is crucial for developing accurate climate models and mitigating the effects of climate change. Solar research helps us comprehend the intricate interplay between solar radiation and atmospheric processes, ultimately enhancing our ability to predict and respond to environmental changes.

CHAPTER 5

Observing the Sun

The study of the Sun has been a pivotal part of astronomy since the dawn of human civilization. Observing the Sun has allowed us to understand more about not only our own star but also the myriad of stars scattered across the cosmos. Since the invention of instruments over 2,000 years ago, we have used various tools to record and study the Sun's light and activity.

With the invention of the telescope, Italian astronomer Galileo Galilei became the first to observe sunspots and solar flares, marking significant discoveries in solar physics. Since Galileo's time, numerous observatories around the world have been established to make regular observations of the Sun. Early spectrographs and photographic techniques paved the way for a deeper understanding of the Sun's internal structure and provided evidence that it is indeed a star. As telescopes and other instruments advanced, we have been able to observe additional features of the Sun, including coronal mass ejections (CMEs) and the complex structure of its magnetic fields. Today, there are many dedicated instruments designed to observe the Sun from both the ground and space.

Observing the Sun is crucial to the field of solar physics. Our familiarity with solar energy and its impact on our environment has driven the study of the Sun for many years. Moreover, the Sun is our only close-up example of a star, offering a unique opportunity to study stellar phenomena in detail. Resolving stars as disks is challenging, requiring large telescopes to collect enough light to create an image. Technological advances made to study the Sun have revolutionized our approach to studying other stars, allowing us to understand just how unique our Sun is.

Historical Observations

Early observations of solar activity significantly contributed to our interest in the Sun and motivated continued study of its essential properties. One of the most popular methods for safely observing the Sun has been the projection or pinhole technique. This method was used by Greek (Theo of Samos, 4th century BC), Arabic (Alhazen, 11th century), and Chinese astronomers (Theophilus, 11th century) to search for transits of Mercury and Venus across the Sun. The motion of sunspots, as well as the phases and transits of planets, were reported in the astronomical records of Aristotle, Tycho Brahe, and Chinese astronomers. Records from this period suggest that the number of sunspots was relatively low between 1650 and 1715, known as the Maunder Minimum.

Determining the precise position of sunspots and measuring the solar diameter led to several important discoveries about the Sun and solar activity. One significant discovery was the variation in the solar diameter and solar energy during the 11-year sunspot cycle, indicating a link between sunspots and solar energy.

There has been considerable progress in observing the Sun over the past few centuries. New data has been acquired not only in the visible range of the spectrum but also in other bands, such as X-rays and radio waves. Early observations, including those in the 16th century, recorded the appearance of dark spots on the Sun. These initial discoveries laid the groundwork for our current understanding, supported by hundreds of years of subsequent observations.

CHAPTER 6

The Sun in Different Cultures

The Sun, while scientifically recognized as the star that our planet orbits, also holds profound significance in various cultural contexts worldwide. It is a celestial body deeply embedded in ancient mythologies and contemporary belief systems, symbolizing life, energy, and divine power.

In ancient Greek mythology, the Sun is personified by the god Helios, who was believed to drive his chariot across the sky each day, bringing light to the world. This personification highlights the Sun's importance in Greek culture as a bringer of life and order.

In the religious practices of indigenous populations of North, Central, and South America, the Sun holds a central place. For instance, the Sun is integral to the agricultural practices of populations in the American Southwest and Central America, reflecting its historical role in sustaining life. The Sun is less frequently personified in Buddhism and Hinduism; however, both religions emphasize harmony with the environment. Early Buddhist texts and images reflect this reverence for nature, while the Indian god Surya, associated with the Sun, plays a supporting role in epics like the "Ramayana."

The Sun has also served as a central figure in globally assembled religions. For much of modern written history, the Sun was used as a cognitive tool to help humans understand their place in the universe. Directions such as North, East, South, and West, and the zenith's erect nature, allowed for navigation and water management, essential for ancient and early historical societies. Understanding the Sun's position and the stars was vital for defining territories and guiding agricultural and navigational practices.

Mythology and Symbolism

The Sun has occupied a special place in mythology throughout human history, often following the Moon in cultural significance. Many cultures have viewed the Sun as a symbol of life, fertility, and renewal. It represents power, happiness, and the highest deity, often considered the first natural phenomenon upon which science was based.

The Sun appears in numerous myths and religions and is often perceived as a symbol of creation and the highest deity. Cultures have created various expressions of the Sun's importance, worshipping it with dances, songs, rituals, and special ceremonies, particularly during festivals and significant moments.

Throughout history, numerous forms of human expression—paintings, signs, statues, and even ancient buildings—have been constructed to honor the Sun. These symbols, arising from local traditions, myths, rituals, and religious beliefs, have become significant cultural elements. For example, in the Basque Country, the Sun's symbol is believed to give power and light. The Sonnenrad, one of the oldest known sun symbols, has been used by pagan tribes and ancient Mesopotamian peoples as a symbol of their spiritual beliefs.

CHAPTER 7

The Future of Solar Research

Numerous new observatories are poised to begin state-of-the-art observations of the Sun, promising to advance our understanding of solar phenomena and their impact on Earth. These cutting-edge facilities and missions will bring unprecedented insights into our closest star.

One of the highly anticipated projects is the **Daniel** mission, set to launch in 2027. This mission will carry the magnetograph Cryo-NIRsp, designed to observe the stellar photospheres of two M-type stars in our near sky with an impressive spectral resolution of about 1 Å in the 10830 Å helium line. Observations at this wavelength are essential for deducing the magnetic topologies of these stars. Simultaneously, the Atacama Large Millimeter Array (ALMA) in northern Chile is preparing to offer unprecedented sensitivity and spatial resolution at millimeter and sub-millimeter wavelengths. This capability will help determine the atmospheric structure linked to the VMC-revealed chromospheric megalobe and its energy source.

In the realm of soft X-ray observations, scientists at the Smithsonian Astrophysical Observatory's High Energy Astrophysics Division have made significant strides. They recently completed work at

NASA's Marshall Space Flight Center in Huntsville, Alabama, integrating the Solar-Terrestrial Relations Observatory (STEREO) A and B solar satellites' command and control subsystems. Since the early 1960s, they have studied X-rays from the Sun's corona, generating high-quality stereo X-ray images of the entire solar corona and its coronal storms. Upcoming advanced coronagraph and stereoscopic telescopes will provide new dynamic information, enhancing computer models of solar wind formation in the inner heliosphere. This work is expected to be completed by the Smith Institute's Blue Sky forecast by June 2024.

Upcoming Missions and Technologies

Several exciting missions set to launch in the coming years will provide new opportunities to advance our understanding of the Sun:

1. **Daniel K. Inouye Solar Telescope (DKIST)**: Built by the National Solar Observatory on the island of Maui, Hawai'i, DKIST is the world's highest spatial resolution solar telescope. In March 2020, DKIST released first-light images with a resolution so fine it could distinguish individual hairs on a person's head from four kilometers away. This telescope will observe the Sun's surface and chromosphere, offering unprecedented details.

2. **European Space Agency's Solar Orbiter**: Launching in 2022 and 2023, the Solar Orbiter mission aims to take high-resolution images of the Sun near its poles, completely outside Earth's orbit. Equipped with remote sensing and in-situ instruments, it will return full-disk and near-Sun images with better than 150 km resolution every week.

New Technologies for Ground-Based Observations

Earth-based telescopes are also receiving technological upgrades to counteract atmospheric distortions. The Mauna Loa Solar Observatory (MLSO) is home to a solar telescope utilizing adaptive optics. This includes the development of Multi-Conjugate Adaptive Optics (MCAO) to correct atmospheric distortions, enhancing our ability to observe photospheric magnetic fields and velocity fields with high spatial resolution. MLSO's home-built F/17.5 Echelle spectrograph will employ MCAO to achieve high-quality spectro-polarimetric observations, providing significant insights into solar dynamics.